Avocados
Eggplants
Breadfruit
Cucumbers
Squashes
Peppers

PLANTS
WE EAT

Tomatoes
Avocados
Eggplants
Breadfruit
Cucumbers
Squashes
Peppers
Tomatoes
Avocados
Eggplants
Breadfruit
Cucumbers
Squashes

Cool as a Cucumber, Hot as a Pepper:

Fruit Vegetables

Meredith Sayles Hughes

Lerner Publications Company/Minneapolis

Check out the author's website at www.foodmuseum.com/hughes

Website address: www.lernerbooks.com

Designers: Steven P. Foley, Sean W. Todd
Editors: Amy M. Boland, Chris Dall
Photo Researchers: Tim Diesch, Kirsten Frickle

LIBRARY OF CONGRESS CATALOGING-IN-PUBLICATION DATA

Hughes, Meredith Sayles.
    Cool as a cucumber, hot as a pepper: fruit vegetables / by Meredith Sayles Hughes.
        p. cm. — (Plants we eat)
    Includes index.
Summary: Discusses the history and uses of plants that are normally thought of as vegetables but share some of the traits of fruits, such as tomatoes, avocados, squash, and peppers. Includes recipes.
    ISBN 0–8225–2832–0 (lib. bdg. : alk. paper)
    1. Vegetables—Juvenile literature. 2. Fruit—Juvenile literature. 3. Cookery (Vegetables)—Juvenile literature. [1. Vegetables. 2. Fruit. 3. Cookery—Vegetables. 4. Cookery—Fruit.] I. Title. II. Series.
SB324.H84 1999
635—dc21   97–28585

Manufactured in the United States of America
1 2 3 4 5 6 – JR – 04 03 02 01 00 99

The glossary on page 85 gives definitions of words shown in **bold type** in the text.

Contents

Introduction

Plants make all life on our planet possible. They provide the oxygen we breathe and the food we eat. Think about a burger and fries. The meat comes from cattle, which eat plants. The fries are potatoes cooked in oil from soybeans, corn, or sunflowers. The burger bun is a wheat product. Ketchup is a mixture of tomatoes, herbs, and corn syrup or the sugar from sugarcane. How about some onions or pickle relish with your burger?

How Plants Make Food

By snatching sunlight, water, and carbon dioxide from the atmosphere and mixing them together—a complex process called **photosynthesis**—green plants create food energy. The raw food energy is called glucose, a simple form of sugar. From this storehouse of glucose, each plant produces fats, carbohydrates, and proteins—the elements that make up the bulk of the foods humans and animals eat.

Sunlight peeks through the branches of a plant-covered tree in a tropical rain forest, where all the elements exist for photosynthesis to take place.

First we eat, then we do everything else.

—M. F. K. Fisher

Plants offer more than just food. They provide the raw materials for making the clothes you're wearing and the paper in books, magazines, and newspapers. Much of what's in your home comes from plants—the furniture, the wallpaper, and even the glue that holds the paper on the wall. Eons ago plants created the gas and oil we put in our cars, buses, and airplanes. Plants even give us the gum we chew.

On the Move

Although we don't think of plants as beings on the move, they have always been pioneers. From their beginnings as algaelike creatures in the sea to their movement onto dry land about 400 million years ago, plants have colonized new territories. Alone on the barren rock of the earliest earth, plants slowly established an environment so rich with food, shelter, and oxygen that some forms of marine life took up residence on dry land. Helped along by birds who scattered seeds far and wide, plants later sped up their travels, moving to cover most of our planet.

Early in human history, when few people lived on the earth, gathering food was everyone's main activity. Small family groups were nomadic, venturing into areas that offered a source of water, shelter, and foods such as fruits, nuts, seeds, and small game animals. After they had eaten up the region's food sources, the family group moved on to another spot. Only when people noticed that food plants were renewable—that the berry bushes would bear fruit again and that grasses gave forth seeds year after year—did family groups begin to settle in any one area for more than a single season.

Organisms that behave like algae—small, rootless plants that live in water

It's a Fact!

The term *photosynthesis* comes from Greek words meaning "putting together with light." This chemical process, which takes place in a plant's leaves, is part of the natural cycle that balances the earth's store of carbon dioxide and oxygen.

Native Americans were the first peoples to plant crops in North America.

Domestication of plants probably began as an accident. Seeds from a wild plant eaten at dinner were tossed onto a trash pile. Later a plant grew there, was eaten, and its seeds were tossed onto the pile. The cycle continued on its own until someone noticed the pattern and repeated it deliberately. Agriculture radically changed human life. From relatively small plots of land, more people could be fed over time, and fewer people were required to hunt and gather food. Diets shifted from a broad range of wild foods to a more limited but more consistent menu built around one main crop such as wheat, corn, cassava, rice, or potatoes. With a stable food supply, the world's population increased and communities grew larger. People had more time on their hands, so they turned to refining their skills at making tools and shelter and to developing writing, pottery, and other crafts.

Plants We Eat

This series examines the wide range of plants people around the world have chosen to eat. You will discover where plants came from, how they were first grown, how they traveled from their original homes, where they have become important, and why. Along the way, each book looks at the impact of certain plants on society and discusses the ways in which these food plants are sown, harvested, processed, and sold. You will also discover that some plants are key characters in exciting high-tech stories. And there are plenty of opportunities to test recipes and to dig into other hands-on activities.

The series Plants We Eat divides food plants into a variety of informal categories. Some plants are prized for their seeds, others for their fruits, and some for their underground roots, tubers, or bulbs. Many plants offer leaves or stalks for good eating. Humans convert some plants into oils and others into beverages or flavorings. In *Cool as a Cucumber, Hot as a Pepper,* we'll take a look at fruit vegetables. Fruit vegetables? What? Let us explain. The edible part of a plant that develops from a flower is defined as a fruit. The fruit of a plant contains the plant's seeds.

While we appreciate the fruit primarily for its taste, its role in nature is to ensure that new plants will grow from the old. Fruits include apples, pears, and oranges. But the category also covers plants we think of first as vegetables, such as tomatoes, avocados, eggplants, breadfruit, cucumbers, squashes, and peppers—all the foods covered in this book. So what's a vegetable? A vegetable is any leafy plant (rather than woody plants, such as trees) that human beings or their animals eat.

You can probably see that the category "fruit vegetable" makes a reasonable slot for the fruitlike foods we eat as vegetables. These are foods that usually appear as an ingredient in a main dish or as a side dish but not often as a dessert or sweet snack.

Many of the fruit vegetables in this book are annuals, which means that after one growing season they die back when the weather cools. If left unpicked, their fruit drops to the ground, where the seeds lie **dormant** through the winter. With the coming of spring, small new plants sprout, ensuring the continuation of another generation.

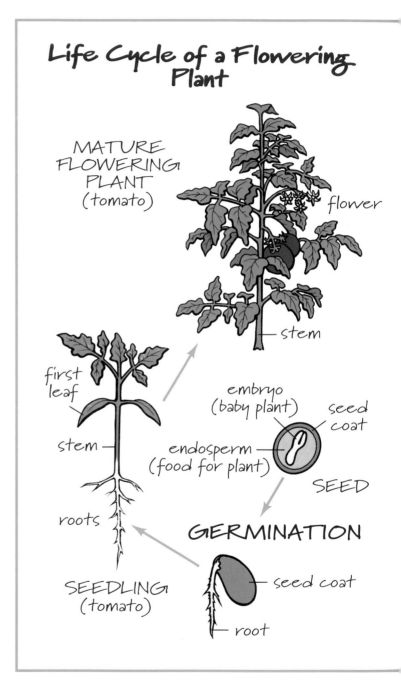

Life Cycle of a Flowering Plant

MATURE FLOWERING PLANT (tomato)

flower

stem

first leaf

stem

roots

SEEDLING (tomato)

embryo (baby plant)

seed coat

endosperm (food for plant)

SEED

GERMINATION

seed coat

root

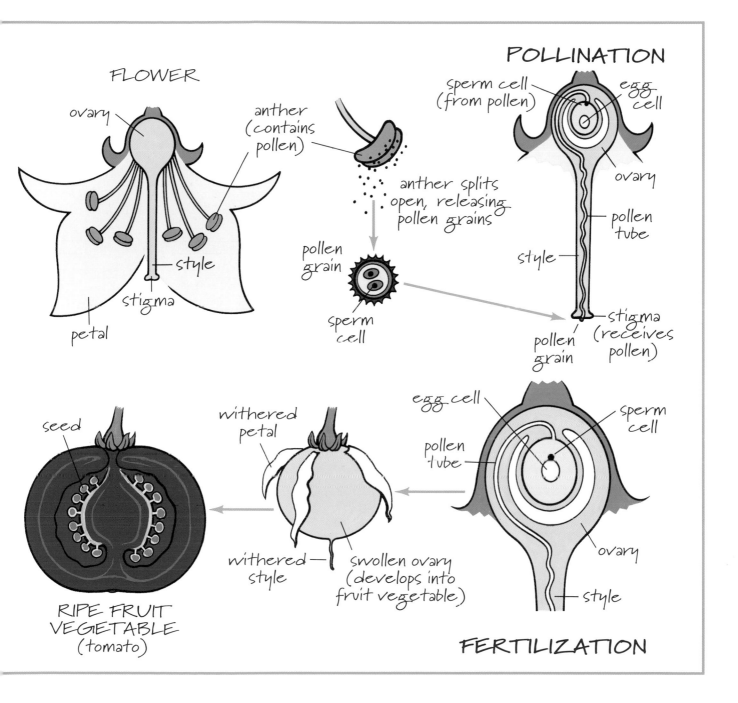

FLOWER

ovary

anther (contains pollen)

style

stigma

petal

anther splits open, releasing pollen grains

pollen grain

sperm cell

POLLINATION

sperm cell (from pollen)

egg cell

ovary

pollen tube

style

stigma (receives pollen)

pollen grain

seed

withered petal

egg cell

sperm cell

pollen tube

withered style

swollen ovary (develops into fruit vegetable)

ovary

style

RIPE FRUIT VEGETABLE (tomato)

FERTILIZATION

Tomatoes

[*Lycopersicon esculentum*]

The tomato was growing wild in South America, probably in the Andes Mountains in Peru, many thousands of years ago. The wild vine produced tiny, short-lived, perishable fruit, and not much of it. The little tomato may have been eaten in season and right on the spot, much as people grab a few wild blackberries growing along a dusty road. It appears that the local Indians didn't cultivate the tomato plant, although they did collect enough of the fruits to eat them with chili peppers. But nowhere does the plant appear either in the pottery or the artwork of the people.

Birds may have carried seeds from the tomato plant north into Mexico, and farmers may have cultivated the plants thereafter. The tomatillo *(Physalis ixocarpa)*, often mistaken for a tomato, was native to Mexico and had been cultivated and used widely by the Indians long before the viny tomato plant. The tomatillo ripens to green in its papery covering on broad festive-looking plants. It thrived in highland regions, while the tomato became a lowland success.

The versatile tomato is one of the world's most popular fruit vegetables.

A world devoid of tomato soup, tomato sauce, tomato ketchup and tomato paste is hard to visualize.

—Elizabeth David

I Say Tomato, and You Say Tomatl

The Aztecs, the native people who established what would later become Mexico City, called the tomatillo *miltomatl* and the tomato *xitomatl* in their language, Nahuatl. If you look closely at both words, you can pick out "tomatl." So did the Spanish explorers who arrived in Aztec lands in the early 1500s. They began calling both the tomato and the tomatillo by the same name, *tomate*.

The Spanish conquerors of Mexico carried the tomato to Europe. The tomato was readily passed to Italy in the 1520s through the Kingdom of Naples, then under Spanish rule. There the Italians experimented with growing a vegetable that responded beautifully to the warm sun. The tomato eventually filled a hole in Italian cooking (after all, who can imagine Italian cooking without tomatoes?).

The first tomatoes to reach Europe were the tiny types known later as cherry tomatoes. Most likely these were egg shaped and yellow in color. The color of the miniature tomatoes determined their Italian name, *pomo d'oro*, or "golden apple."

In 1544 the Italian herbalist Pierandrea Mattioli wrote the first account of the tomato, which people ate raw with oil and salt. His opinion was fairly negative, perhaps because someone he knew may have become sick after eating the leaves or stems of the plant by mistake. The tomato's greenery is

It's a Fact!

To the Aztecs, the word *tomatl* meant "something round and plump." They added the prefix *xi* to better describe the tomato plant. *Xitomatl* means "plump thing with a navel." (Look at a tomato and you'll see the navel, the point where it once connected to the mother plant.)

Pierandrea Mattioli

Tomatoes come in a wide variety of shapes, sizes, and colors.

Love Apple, Anyone?

It may have all started when Italians either confused or associated the tomato with its cousin the eggplant, then known as *pomo del moro*, or "apple of the Moors." The French term *pomme des Mours* sloppily became *pomme d'amour* (which sounds similar). *Amour* means "love" in French, and there you have it, love apple. Some tomatoes do vaguely resemble the human heart, but the name is just a pun made from French words having absolutely nothing to do with love, Saint Valentine, or anything mushy.

indeed highly toxic. Despite Mattioli's bad review, the tomato was first cultivated as a food plant by Italians, even though the rest of Europe grew it purely as a viny ornamental oddity.

The Moors, an Islamic group originally from North Africa, probably introduced the tomato to the Middle East in the 1500s, after meeting up with the plant in southern Spain, which they had ruled earlier. The tomato then made its way to Asia with Dutch traders, who carried the plant to Indonesia as early as 1658. The British most likely brought tomatoes to India at about the same time.

European colonists brought tomatoes to North America in the early 1700s, but the plant was not mentioned formally in print until the late 1700s. Tomatoes were reported to be growing in Japan by the early 1800s.

Workers and machines unload bushels of tomatoes at the H.J. Heinz plant in Fremont, Ohio, in 1945.

How Do Your Tomatoes Grow?

Home gardeners in North America were growing tomatoes long before the plant became a commercial statistic. It probably truly took off during World War I (1914–1918), when the U.S. government urged citizens to grow food for their families.

Tomatoes are surprisingly easy to grow. Young plants are placed in well-plowed, composted soil. Because tomato plants tend to spread up and out, they are usually put in the ground with a wooden stake to support the fruit-laden offshoots. Although tomatoes require consistent levels of water when first planted, they flourish with less water after the fruits begin to form. After a healthy tomato plant starts producing, it simply requires appreciative tomato lovers to keep its fruit picked. Many tomato plants keep on going right up until the first frost.

The tomato is the most commonly grown U.S. garden vegetable—and for good reason. Taste! Comparing a supermarket tomato to a garden-grown tomato is a no-contest affair.

Family Matters

To keep things straight in the huge families of plants and animals, scientists classify and name living things by grouping them according to shared features within each of seven major categories. The categories are kingdom, division or phylum, class, order, family, genus, and species. Species share the most features in common, while members of a kingdom or division share far fewer traits. This system of scientific classification and naming is called taxonomy. Scientists refer to plants and animals by a two-part Latin or Greek term made up of the genus and the species name. The genus name comes first, followed by the species name. Look at the tomato's taxonomic name on page 10. Can you figure out to what genus the tomato belongs? And to what species?

Growing and Selling Everybody's Favorite Fruit Vegetable

A year-round supermarket tomato is lower in quality than a just-picked summer tomato because the store-bought kind are grown to withstand commercial handling and to travel well. Commercial tomato growers use both machines and hand labor to bring their crop to consumers.

Growers plant seeds in prepared greenhouse beds. When each plant is about nine inches tall from its roots to its leaves, the tomatoes are ready for transplanting into outdoor raised beds.

At one large farm, six people sit in seats on the transplanter machine, which moves slowly through three rows at a time. The workers

A young, green tomato grows on the vine. Growers take the tomato seedlings outdoors after four to six weeks.

place the young tomato plants on a turning wheel that lowers each plant to the ground. The transplanter's blades push the soil around each plant as the machine moves along.

Growers water the tomatoes by drip irrigation or furrow irrigation. In the drip method, thin pipes below the soil put a slow but steady amount of water and fertilizer into the ground. With furrow irrigation, farmers periodically release water from canals into the rows of plants. Tomatoes grown in the eastern United States often rely solely on rainfall.

It's a Fact!

Farmers in Tennessee have been experimenting with growing tomatoes on untilled soil. Plowed fields can erode in heavy rain or wind, causing the topsoil to run off into lakes. The tomato plants are placed in holes dug right through a turf of rye or other grass. The grass surrounding the plant provides the tomato with mulch and nutrients and also keeps weeds at bay.

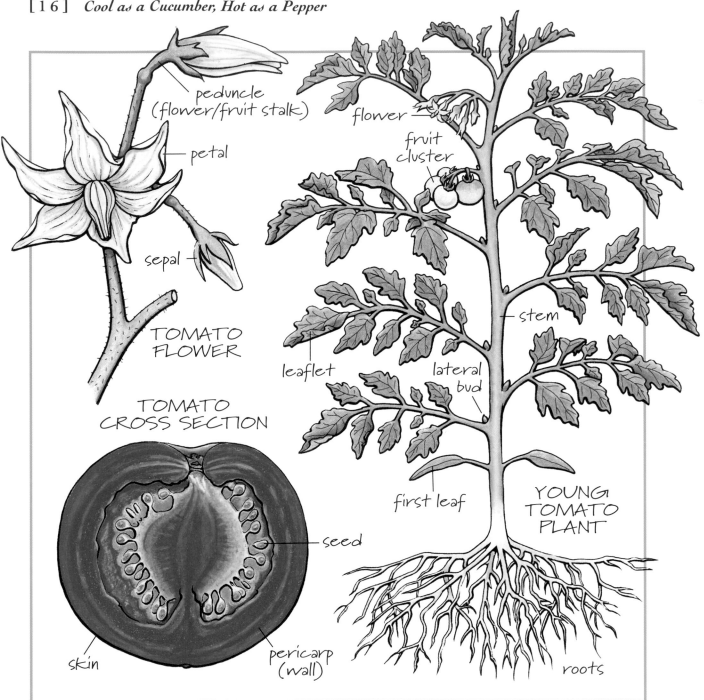

peduncle
(flower/fruit stalk)

petal

sepal

TOMATO
FLOWER

flower

fruit
cluster

stem

leaflet

lateral
bud

first leaf

YOUNG
TOMATO
PLANT

TOMATO
CROSS SECTION

seed

skin

pericarp
(wall)

roots

After about 90 days, the tomatoes are ready for picking. Commercial tomatoes to be sold fresh are picked entirely by hand, usually when they are still green. Crews of 25 or so people walk through the rows, plucking the tomatoes and placing them in plastic buckets. The pickers empty their buckets into a trailer that carries the fruit to other workers who sort, size, grade, and box the tomatoes. Laborers stack the boxes of tomatoes on pallets and bring them into ripening rooms for as long as two days. In these rooms, the green tomatoes are sprayed with ethylene gas, a chemical that occurs naturally in fruits. The gas encourages the tomato to redden but not actually to ripen. Then the tomatoes are loaded onto temperature-controlled trucks and transported to supermarkets throughout the United States.

Commercial growers who pick their tomatoes green say that the product they bring to the store is well worth eating. But often, for the sake of convenience, supermarkets store the tomatoes in coolers with other vegetables. That's too bad, because tomatoes should never be refrigerated! Below 55 degrees, their taste is destroyed.

By contrast, commercial tomatoes for processing are allowed to ripen on the vine and are picked by machine. These tomatoes can afford to suffer bruises—they don't have to look perfect to be smashed into cans or chopped up into sauces. The tomato harvester moves through the fields, cutting each plant from its roots and pulling the plants up from the ground. The machine then shakes the tomatoes off the stalks and onto a conveyor belt. The belt moves the tomatoes along so workers can hand sort them. A trailer truck brings the tomatoes directly to the processing plant.

Commercial growers are looking for ways to improve tomatoes so they can move the plant more quickly to market and also ensure a tomato that more closely resembles the taste and flavor of vine-ripened tomatoes.

Hydroponic (grown in water without soil) tomatoes are beginning to make inroads around the world during winter months.

A mechanical harvester operates in a tomato field in California.

Air Tomatoes

In Japan growers are raising tomato plants as tall as trees in greenhouses that use solar power to generate light 24 hours a day. These growers rely on an advanced method called aeroponics to raise tomatoes. In this method, tomato plants grow without soil in enclosed air chambers that are periodically sprayed with a mix of water, nutrients, and growth hormones.

In the United States, aeroponics helps to start new plants to get them ready for transplanting quickly, either into soil or into hydroponic settings. Growers coined a new name for these modern growing methods—agribiotechnology.

Supplied by Mexico, the Netherlands, Canada, Israel, and Belgium, these tomatoes are grown in greenhouses and fed nutrients through water. They are allowed to ripen almost completely and are either packed individually or as a stem bearing several tomatoes. Hydroponic tomatoes cost more, but the consumer is getting a tasty tomato at a time of year when tomatoes aren't normally available.

Tomatoes Are Big Business

Even though commercial tomatoes don't taste as good as homegrown, U.S. consumers buy more than $4 billion worth each year. In 1996 Mexico supplied the United States with 1.5 billion pounds of fresh tomatoes for sale in grocery store produce departments. In the same year, Florida grew 1.2 billion pounds and California raised 1 billion pounds of fresh tomatoes.

Hydroponic tomatoes grow on supports called trellises, which keep the vines from spreading on the ground.

The market for processed tomato products, such as soups and sauces, is huge. People in the United States eat slightly more than 74 pounds of processed tomatoes per person in a year.

In the processed tomato business, California is the world's leader. The state's growers raised a phenomenal 23 billion pounds of processing tomatoes in 1996, all turned into salsa, pasta sauces, juices, pizza sauce, ketchup, tomato paste, barbecue sauce, canned tomatoes, and many other processed foods.

Pomo d'Oro with Everything, Please

Consider the crisp Naples-style pizza with tomato sauce, thin slices of cheese, fresh basil, and olive oil. Or think about plump cheese ravioli with a rich marinara sauce, fresh tomatoes stuffed with herbed rice, spaghetti with *bolognese* sauce, and cannelloni filled with spinach and cheese, all baked under a warm blanket of bubbly tomato sauce. These are just a few of the Italian tomato-based dishes the world appreciates.

The tomato found a happy home in the dry, sunny soils of southern Italy, where people still hang bunches of cherry-sized tomatoes on the sides of their houses to dry in the sun. But it also enjoyed good growing

It's Official

Back in 1893, the tomato was officially declared by the U.S. Supreme Court to be a vegetable. The question had been hotly debated and arose because importers of tomatoes were annoyed at having to pay a tariff (tax) on all imported vegetables. Since there was no tariff on fruits, the importers attempted to have the tomato declared a fruit. Associate Justice Horace Gray wrote, "Botanically speaking, tomatoes are the fruit of a vine, just as are cucumbers, squashes, beans and peas. But in common language all these are vegetables, which are usually served at dinner with the principal part of the repast and not, like fruits generally, as a dessert."

Eugenie, the Spanish wife of the French emperor Napoleon III, has been credited with introducing tomato dishes to France in the mid-1850s.

It's a Fact!

The Campbell Soup Company, which started out selling asparagus in Camden, New Jersey, created the first successful line of condensed soups in 1897. Its lead offering was tomato soup created from bushels of beefsteak tomatoes grown in the state's sandy soils. The earliest Campbell soup can label featured two men struggling to carry a jeep-sized tomato hung from a long pole.

conditions in Spain, where it first landed on its arrival from Peru. Gazpacho, a cold soup whose primary ingredient is tomatoes, is an original from Seville, Spain. Ingredients in the soup also include other fresh vegetables, olive oil, water, and finely ground bread soaked in wine vinegar. In fact, the word *gazpacho* comes from an Arabic term meaning "soaked bread."

The tomato found favor in France, a country never willing to miss a culinary treat. The region of Provence in southern France has given the world many dishes *à la provençale,* (or, "Provence style"), most of which include tomatoes. The most famous of these is *tomates à la provençale,* a simple, never-fail way to eat tomatoes. You can prepare this dish yourself with the recipe on page 23.

As soon as the tomato landed in the Middle East, people immediately paired the vegetable with its Asian cousin the eggplant. A stew of baked eggplant, tomatoes, and chickpeas is popular in many Middle Eastern countries. Tomatoes grilled on skewers with swordfish or lamb is a Turkish specialty. Greek salad with feta cheese almost always includes pieces of tomato as does *tabbouleh,* the crushed wheat and mint salad of Lebanon. Grilled tomatoes or tomatoes stuffed with rice accompany many meals in Iran, Iraq, and Egypt.

Tomatoes are eaten in most parts of Africa, too. In Kebili, Tunisia, local bakers make a flatbread called *khubs m'tabga* from tomatoes, chili peppers, flour, and green onions. The bread is often eaten with a tomato stew in which eggs are poached. Malay people in South Africa make a layered meat stew called *bredie* with many vegetables. The tomato version, always made with chilies, is a favorite.

Tomatoes, whether cooked or served fresh, are a delicious and colorful addition to any meal.

Move Over, Ketchup

Aside from soups, juices, and sauces, what other tomato-based essential comes to mind? Ketchup, did you say? Or was it salsa, the condiment that has gained steadily on ketchup since the late 1980s? The consumer shift from ketchup to salsa in much of North America not only salutes the tomato's versatility but shows the growing influence of Hispanic culture throughout the continent.

Ketchup originated in China, where it was known as *ke-tsiap*. The soy and fish-based sauce so pleased British sailors in the area in the 1600s that they brought the product back to Britain with them. At first the sauce was made with anchovies and nuts, but these were shoved aside when the tomato crept into the mix in the late 1700s. An American named James Mease, then living in New

To Your Health!

In the United States, people eat so many tomatoes that these fruit veggies are our biggest supply of vitamins and minerals, even though tomatoes are only the fourteenth most nutritious vegetable in the national diet. It seems we would rather eat tomatoes than the 13 more nourishing choices—broccoli, spinach, Brussels sprouts, lima beans, peas, asparagus, artichokes, cauliflower, sweet potatoes, carrots, sweet corn, potatoes, or cabbage.

Salsa's versatility has made it more popular than ketchup. In 1996 U.S. retail sales of salsa were about $690 million, compared to $456 million for ketchup.

Jersey, concocted tomato "catchup" but didn't publish his recipe until 1812. The new recipe combined cinnamon, allspice, and dry mustard with vinegar and, of course, tomatoes. By the mid-1800s, Americans were eating commercially prepared ketchup with gusto. When the H. J. Heinz Company of Pittsburgh brought out its ketchup in 1876, it zoomed to the top of the commercial list.

As for salsa, no such timetable exists for a food that came out of Mexican kitchens and traveled northward into the southwestern United States and eventually throughout North America. Traditional salsa is a sauce made from chopped tomatoes, green onions, cilantro (an herb), chilies, salt, and pepper. It is served as a dip with tortilla chips, as a topping for tacos and burritos, and as an accompaniment to chicken and fish dishes. Foods similar to salsa are also found in Asia, Africa, and the Middle East.

Dig In!

TOMATES À LA PROVENÇALE
(2-4 servings)

2 large tomatoes, not too ripe
½ cup bread crumbs
1 tablespoon chopped parsley
3 garlic cloves, crushed and finely chopped
dash of salt and pepper
olive oil, enough to lightly cover bottom of pan

Cut the tomatoes exactly in half, from top to bottom. Remove the seeds with the tip of a small knife. Coat the bottom of a large skillet with olive oil. Cook the tomatoes face down over medium heat for a few minutes until the sides begin to curl up. Turn them over and cook two or three minutes more, shaking the pan or adding a bit more oil if the tomatoes begin to stick. Then place the tomatoes face up in a shallow casserole and lightly salt and pepper them. Add a bit more oil to the skillet, and cook the bread crumbs, chopped parsley, and garlic, stirring often, until the garlic is soft. Then sprinkle the mixture on top of the tomato slices. Bake in a hot oven (400°) for three minutes.

Tomates à la provençale make a good side dish or, with bread, even a light meal.

Avocados
[*Persea americana*]

Say *ahuacatl* (ah-wah-KOT-ll). Sounds something vaguely like avocado, doesn't it? It should—it's the word the ancient Aztecs used for this oil-rich, protein-filled fruit veggie. Originally from Mexico, the avocado's wild relatives date as far back as 9,000 years. Archaeologists in Peru have found avocado pits buried with 3,000-year-old mummies. Scientists think avocados also were growing early on in Colombia, Ecuador, and Guatemala.

Seeking New Tropics

The avocado's travels beyond Mexico and South America began with the Spanish conquistadores (conquerors) who arrived in Mexico in 1519. Within five years, the Spanish had enslaved the majority of the native Aztec peoples living in and around Tenochtitlán, a wondrous lakeside city that the Spanish destroyed to build what would become Mexico City.

The fruit of the avocado tree

> . . . *[the avocado's] unctuous oily flesh, a trait shared only with the olive and the coconut in the edible plant world, intrigued the natural history writers enough to be given long chapters in their books on the vegetable wonders of the New World.*
>
> —Sophie Coe

A young Guatemalan proudly displays his avocados. Guatemala is one of the Central American countries where avocados are grown extensively.

It's a Fact!

Jamaicans are credited with giving the avocado the nickname "alligator pear." The English-speaking people of this West Indian island—once a Spanish colony—substituted the word "alligator" for aguacate, the Spanish term for "avocado," and added "pear" for the fruit's pearlike shape. The commercial avocado industry discourages the use of this name, perhaps fearing that consumers won't want to buy or eat something associated with a 'gator.

The Spanish observed that the Aztecs mashed the avocado, maybe with wild onions and tomatoes, and ate the spread with corn tortillas. The conquistadores and other foreign adventurers who sampled the avocado in the region also tried it with salt and sugar. In 1527, three years after their conquest of Mexico, the Spanish carried the avocado back to Spain.

Later on, in the 1700s and 1800s, British sailors bought avocados in Caribbean ports to make "midshipman's butter," a diplike concoction that went well with hard shipboard biscuits. Future U.S. president George Washington, then 19 years old, accompanied his ailing brother Lawrence to Barbados in 1752 and there sampled *agovado* along with pineapple and grapefruit.

History is obscure as to how, when, and with whom avocados made their way around the world. We do know that in 1833 Henry Perrine planted the first domesticated avocado in Florida. Florida has since become an avocado paradise. Farmers in the state grow about 56 different avocado varieties. Dade County, southwest of Miami, is Florida's avocado belt. Farms in the area supply about 5 percent of the total U.S. avocado crop.

Avocado Portrait

A relative of the laurel tree, the avocado has spicy relatives in cinnamon, bay, and sassafras trees. Mexican cooks bake some foods wrapped in the aromatic leaves of certain avocado varieties. A tree of the **tropics,** the avocado is grown extensively in Mexico, Central and South America, the West Indies (an island region in the Caribbean Sea), Israel, South Africa, California, Hawaii, and Florida. The tree also turns up in Australia, India, North Africa, and a range of Pacific islands.

The pear-shaped, green to black fruit vegetables hang on an evergreen that can grow as tall as 80 feet well into maturity. Its glossy dark green leaves drop from the tree in early spring, but enough leaves remain in place to assure the tree's evergreen status. Because of the heavy leaf litter, nothing else will grow under a tall avocado.

Avocados are generally divided into three main family groupings—the West Indian, the Guatemalan, and the Mexican. Thin-skinned West Indian varieties are often grown in Florida, while hardier Mexican examples are more common in California. The famous bumpy-skinned Haas avocado is a Guatemalan variety. California grows 95 percent of all the avocados raised in the United States. Farmers there cultivate seven different varieties, although the Haas variety accounts for most of the volume.

The hard pit of the avocado is surrounded by a papery husk.

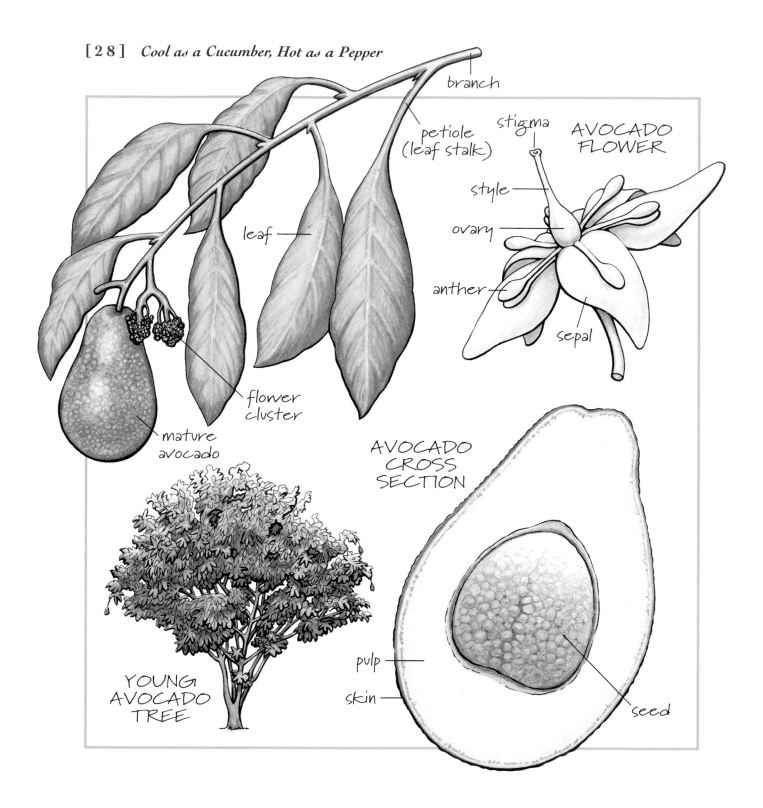

branch

petiole
(leaf stalk)

stigma

AVOCADO
FLOWER

style

ovary

anther

sepal

leaf

flower
cluster

mature
avocado

AVOCADO
CROSS
SECTION

YOUNG
AVOCADO
TREE

pulp

skin

seed

At this orchard in Santa Barbara, California, avocado trees are planted in rows about 25 to 30 feet apart.

Commercial growers plant avocado trees about 15 to 25 feet apart, with 25 to 30 feet between rows. In California workers still harvest avocados by hand, using shears called clippers attached to poles up to 14 feet long. The pickers place the avocados in nylon bags that can hold about 40 pounds of fruit. (One tree can produce as many as 400 avocados a year.) Workers then put the avocados in large picking bins, which hold between 600 and 800 pounds of avocados. Tractors haul the bins to transport trucks that carry as many as 24 bins at one time.

At the packing shed, the bins are immediately placed in cold storage for 24 hours. After they are cooled, the avocados are ready for sorting and packing. The bins are carefully tipped over on a conveyor belt, where they are sorted by size. Next the fruit is brushed, washed, and packed 48 avocados to a lug (carton). The lugs are then stacked on pallets, where refrigerated trucks pick them up for delivery all across the country.

Cooling down the fruit helps to keep it from overripening.

Dig In!

Save the pit of a really ripe avocado (it will probably sprout faster). Remove the papery husk from the pit. You can grow the avocado either in water or in soil. If you choose water, stick three or four toothpicks about halfway down the pit. Perch the pit on the rim of a large jar filled with enough water so that the bottom, rounded half of the pit is under water and the pointed end is facing up.

After a couple of weeks, you'll see roots growing in the water. When a sprout appears at the top, transplant the avocado into a two-gallon planter filled with potting soil.

A pit planted in soil from the very beginning will grow faster. Push the pit into the soil mix until the pointed end sticks out about one inch. Keep soil evenly moist. Either way you choose to grow the pit, you'll eventually end up with an avocado tree—and we mean tree! But you won't grow any avocados indoors. Sorry.

The biggest market for California avocados is the greater Los Angeles area, which isn't far from San Diego County—the state's avocado center. Growers in this county raise more than 320 million pounds of avocados each year. Southern California's warm climate allows year-round production.

Spain, Mexico, and Israel are all major producers of avocados, too, with Kenya, Brazil, Argentina, and South Africa also vying for their share of the market. France buys the bulk of Israel's avocado export harvest, with Germany and the Scandinavian countries ranking as the next best customers. As competition heats up in European markets, Israeli growers are looking to reach out to Japan and other countries of

La Habra, a suburb of Los Angeles, is the birthplace of the Haas variety of avocado.

To Your Health!

Good fat, bad fat, fat fat. You've probably heard lots of talk about fat. Avocados are rich in fat for a plant product. The fat in avocados is good fat—the kind that tends to break down cholesterol, which can form an unhealthy layer on the insides of our blood vessels. But fat is still fat, and nobody should eat too much, no matter what type.

Avocados also contain significant amounts of protein and whopping amounts of potassium, far more than other fruits and vegetables. Potassium helps build muscles and regulates the heartbeat. If you feel exhausted and you're achy after a sweaty workout, you probably need a boost of potassium. Grab an avocado, and you'll also be getting vitamin B_6 and vitamin A!

the Pacific Rim as their newest avocado customers.

It's Everywhere

Japanese sushi rolls contain avocados, as does Brazilian ice cream. Avocados go into hot soups made in Colombia and Ecuador and into a Jamaican cold soup flavored with fresh lime juice and chili peppers. Nicaraguans bake avocados, and Indonesians mix the fruit vegetable with milk, coffee, and rum for a potent cold drink.

Seviche—a raw fish, lime juice, and avocado appetizer—is a favorite in Peru. In Mexican and New Mexican cooking, avocados turn up in a smooth dip called guacamole and in several tasty tortilla combinations.

High Technology

Many farmers rely on **pesticides** to kill the insects that attack their crops. Some avocado growers use alternative methods of insect control. For example, growers in Queensland, Australia, have had success relying on a combination of spiders and ladybugs to keep unwanted insects at bay.

Using insects to control other insects has a fancy name—integrated pest management. This means introducing into fields those insects that eat the insects that eat avocados. As a result, avocados are apt to carry much lower levels of pesticides—which can harm people, too—than other foods.

It's a Fact!

Did you know that some kids in Homestead, Florida, attend a grade school called Avocado Elementary?

Other methods of integrated pest management include the selection of pest-resistant plants and crop rotation, which discourages insects from preying on a field from year to year.

Dig In!

GUACAMOLE
(2 Cups)

2 avocados, slightly soft to the touch
½ cup prepared salsa, medium or mild
the juice of half a lemon or lime

You may want to ask an adult to help you open the avocados. For each avocado, insert a paring knife until you reach the hard pit in the center. Cut the avocado in half the long way around the pit. Grasp the avocado on each side of the cut and twist until one half comes off the pit. Take the part that still holds the pit and cut *that* in half, slicing around the pit from top to bottom. Repeat the twisting maneuver. Pull the pit out of the last small piece of avocado—careful! It's slippery. Scoop out the pulp from the avocado pieces and place in a wide bowl. Mash the avocados with a fork. Then add the salsa and lemon or lime juice. Mix well and eat at once with tortilla chips or bread.

If you like to mash things, this delicious dip from Mexico is for you.

Eggplants
[*Solanum melongena*]

Shining in shades of lilac, deep purple, and poster-paint white, the eggplant is a beautiful food to look at. It grows on a plant with flowers resembling those of the tomato and the potato, and its deep green leaves are traced with purple veins. The eggplant thrives in heat and succumbs to chilly temperatures and frost.

Although much admired by many cooks, the eggplant does not hold the same place of importance in world cuisine as its cousins the pepper, tomato, and potato. All are members of the Solanaceae (nightshade) family. Once believed to be highly poisonous, the eggplant and its cousins are safe to eat as long as you're eating the right part of the plant—the fruit! The leaves of Solanaceae plants are, indeed, highly toxic.

Although not as popular as the other members of the Solanaceae family, the eggplant is an important element in Mediterranean and Middle Eastern cooking.

. . . these apples [eggplants] have a mischievous quality.

—John Gerard

The heirloom garden at Monticello

The only member of the Solanaceae family to come from the Eastern Hemisphere, the eggplant was probably first cultivated in India about 4,000 years ago. Indian farmers may have developed the plant from spiky varieties with tiny, egg-shaped fruits. Transported by Arab traders from India, the eggplant arrived in Spain during the 1100s and soon settled in the lands surrounding the Mediterranean Sea. In the 1400s, Italians were growing eggplants, which reached England during the next century. Louis XIV, the king of France, had a passion for foreign vegetables. In the late 1600s, he brought the eggplant to France and had his gardeners raise it in his own kitchen garden.

U.S. president Thomas Jefferson, an experimental gardener, is credited with bringing the eggplant to North America. It is likely that he obtained eggplant seeds or cuttings from France at the end of the 1700s. Jefferson's garden records indicate planting eggplant in July of 1809. According to his brief notation, the effort failed. Other plantings must have flourished because a white, prickly variety of eggplant is still grown in the **heirloom garden** at Monticello—Jefferson's Virginia estate.

The Chinese eggplant *(left)* is a smaller and thinner version of the standard eggplant. This young eggplant *(below)* is about to be planted.

Planting and Harvesting "Eggs"

As with many fruit vegetables, growing and harvesting eggplant is done almost entirely by hand. One large Florida operation hires about 250 farmworkers to cut eggplant during the season.

Started from seed in nursery beds, the small plants are transplanted when they are between six and eight inches high. Workers have broken up and fertilized the soil in advance. A tractor rolls through the field with a hole punch, marking spaces about 18 inches apart into which the transplants will go. Crews carry trays holding about 80 to 100 seedlings into the rows, place the plants into the ground by hand, and cover up the roots with soil. Often growers cover the rows with plastic sheeting to keep the plants warm through cold weather. Eggplants are frequently supported with stakes, much as we stake tomatoes in home gardens.

During the growing cycle, machines roll through the fields spraying pesticides, but come harvest time, the work is done by hand. Eggplants have stiff thorns near the stem, so workers wear gloves to protect themselves. They cut the eggplants with garden shears and place them into buckets for later collection in bin boxes. Pickers harvest each row two to three times per week, selecting only eggplants of the right size.

Trucks collect the bin boxes and transport them to the packing house, where the eggplants are dumped onto conveyor belts to be washed, sized, and graded. Line workers wrap fancy-style eggplants individually in paper and pack them in cartons. "Fancy" means well shaped, well colored, undamaged eggplants. Bulk eggplants—ungraded produce of varying shape, size, and quality—go into the boxes as they are. While eggplant is sometimes cooled down before packing to help prevent spoilage, the produce is frequently processed and packed right in the field and then trucked directly to supermarket distribution centers.

At this nursery in southern Lebanon, a worker sprays eggplants with pesticide to protect against a type of beetle. The bugs eat holes in the leaves, which can harm the fruits and prevent successful growth.

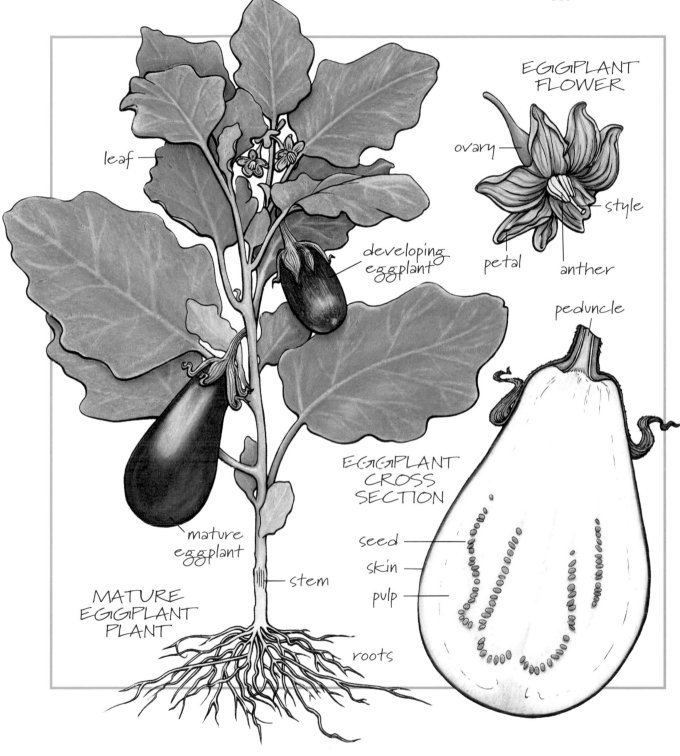

EGGPLANT
FLOWER

ovary

style

petal

anther

leaf

developing
eggplant

peduncle

EGGPLANT
CROSS
SECTION

seed

skin

pulp

mature
eggplant

stem

MATURE
EGGPLANT
PLANT

roots

Dig In!

BABA GANOUCHE
(2 Cups)

1 large eggplant
Juice of 2 lemons (about 6 tablespoons)
4 tablespoons tahini, thinned according to directions on can
3 cloves garlic
1 teaspoons salt
4 tablespoons chopped fresh parsley for garnish

Preheat oven to 400°. Prick the eggplant several times with a fork. Place it on the oven rack and bake for 30 minutes or until very soft. Remove it from the oven and cool. When the eggplant is cool enough to handle, peel off the skin. Cut the eggplant into chunks, place them in a medium bowl, and mash with a potato masher or puree in a food processor until smooth. Beat in the lemon juice and tahini. In a small bowl, crush the garlic into the salt with the back of a spoon. Stir the garlic and salt into the eggplant mixture. Put the mixture in a serving dish and garnish with the parsley.

According to one story, a cook created baba ganouche as a soft, yet delicious dish for his aging, toothless father. The one unusual ingredient is tahini, a paste of ground sesame seeds. Look for tahini at a natural foods, Asian, or Middle Eastern grocery store.

It's a Fact!

It has been said that the Turks cook eggplant 40 different ways and have a thousand eggplant recipes.

Eggplant parmesan

Experiencing Eggplant

From Spain all the way to the Middle East, cooks in countries bordering the Mediterranean prepare eggplant with garlic and olive oil in appetizing variations. Southern Italy has eggplant parmesan, a world-famous dish of layered fried eggplant and tomato sauce with cheese. Less well known is eggplant stuffed with anchovies, capers, and olives. The ratatouille of southern France combines eggplants, tomatoes, onions, and olive oil into a robust vegetable stew that can be eaten hot or cold with French bread.

People in countries that border the eastern coast of the Mediterranean Sea cook eggplant daily. Turks stew eggplant with onion and tomato or slice and grill the vegetable with olive oil. Greek moussaka is a layered dish using eggplants, meat, potatoes, and tomatoes. Broiled and then mashed with lemon juice and garlic, eggplant becomes *baba ganouche,* a popular dip throughout the Middle East. Egyptian cooks stuff eggplants with dill-flavored rice for a baked dish called *mehshi.*

In India eggplant dishes call for a wide variety of spices. Cardamom, cinnamon, turmeric, coconut, and coriander leaves are frequent associates of eggplant. People in China and Thailand make spicy eggplant dishes with hot chili peppers.

Breadfruit

[*Artocarpus communis*]

Probably native to the Malay Archipelago, breadfruit either drifted on the sea or was carried by early peoples to the Pacific Islands well before written history. The plant has been cultivated in that part of the world for thousands of years.

The first European to write about breadfruit was the British explorer William Dampier, who saw the breadfruit tree for the first time on the Pacific island of Guam in 1688. There and in Samoa (a group of Pacific islands), breadfruits were baked with hot stones in pits dug into the ground. The wood of the trees—which grew as high as 60 feet—was also used for canoes, and the bark was made into cloth. In Hawaii the wood of the breadfruit tree was prized for making surfboards and drums.

Grown mostly in tropical areas, breadfruit has a starchy pulp that some people think feels like bread.

[The taste of breadfruit is] *as disagreeable as that of a pickled olive generally is the first time it's eaten.*

—Captain James Cook

A bountiful plant, one breadfruit tree can supposedly feed a large family for one year. Breadfruit caught the attention of British explorer Captain James Cook in the late 1770s, though his description of the plant's taste was less than enticing. Others in the British navy found the breadfruit acceptable, in part because they were relieved to have fresh food after a long journey at sea.

An obsession with breadfruit fueled one of the most famous mutinies in history. In 1787 Britain's King George III sent Captain William Bligh to carry breadfruit trees from Tahiti in the South Pacific to the West Indies, where British plantation owners wanted to grow the plant as a cheap **staple food** for feeding the African slaves who worked the sugarcane fields.

On the trip to the Caribbean from Tahiti, the deck of the H.M.S. *Bounty* was crowded with a thousand young breadfruit plants. Fresh drinking water was scarce, but Captain Bligh (later known as "Breadfruit Bligh") insisted most of it be used to water the breadfruit. The thirsty men, already angered by the captain's tyrannical ways,

revolted. They tossed the plants overboard and lowered the captain and a few of his faithful followers over the side into a small boat. Although that shipment of breadfruit plants never reached its destination, a second shipload did—under the command of Captain Bligh. In 1793 Bligh arrived in Jamaica with the promised cargo, and in a matter of years, the breadfruit tree was a common sight throughout the West Indies.

Growing and Selling

Growers raise breadfruit either from root cuttings or from seed. The young plants need ample rain and well drained soils. Once mature the trees will produce two harvests each year. The fruit itself has a knobby outer skin and is large, about eight inches in diameter. Workers climb breadfruit trees to harvest the fruit before they are fully ripe. Cut with a sharp knife, the fruit is carried carefully back down to the ground in sacks. Workers then place the fruit stem down to allow its bitter latex (sap) to run onto the ground. Sometimes harvesters reach up from the ground with long pruning poles to dislodge the fruits, catching them in baskets or bags as they fall.

It's a Fact!

Caribbean islanders not only eat breadfruit, they make tea from it as a cure for high blood pressure. In addition, they caulk their boats with a gummy substance from the plant and weave cloth from the bark's inner fiber.

(*Facing Page*) After overthrowing Captain Bligh, the mutineers sailed the *Bounty* to Pitcairn Island in the southern Pacific Ocean. They ultimately burned the ship on the beach, because returning to Britain would have meant death for them all. Descendants of the British sailors and their Tahitian wives still live on Pitcairn Island.

The seeds of the breadfruit, sometimes called breadnuts, are also edible.

After the latex has drained off, workers wash the fruit and pack it in cartons that go to cold storage to cool the fruit down for shipment to market. Breadfruit sent by air is picked no more than one day before shipping because the fruit deteriorates quickly. Breadfruit is eaten before it is ripe, so if the fruit begins to soften, it can no longer be sold. Although breadfruit markets outside the Caribbean are limited, the fruit is exported in small quantities from Barbados, Grenada, St. Lucia, Dominica, and St. Vincent to Canada, Britain, the United States, and the Netherlands.

It's a Fact!

A saying among peoples of the Pacific Islands goes something like this, "No breadfruit can be reached when the picking stick is too short." Translation—preparation is the key to success.

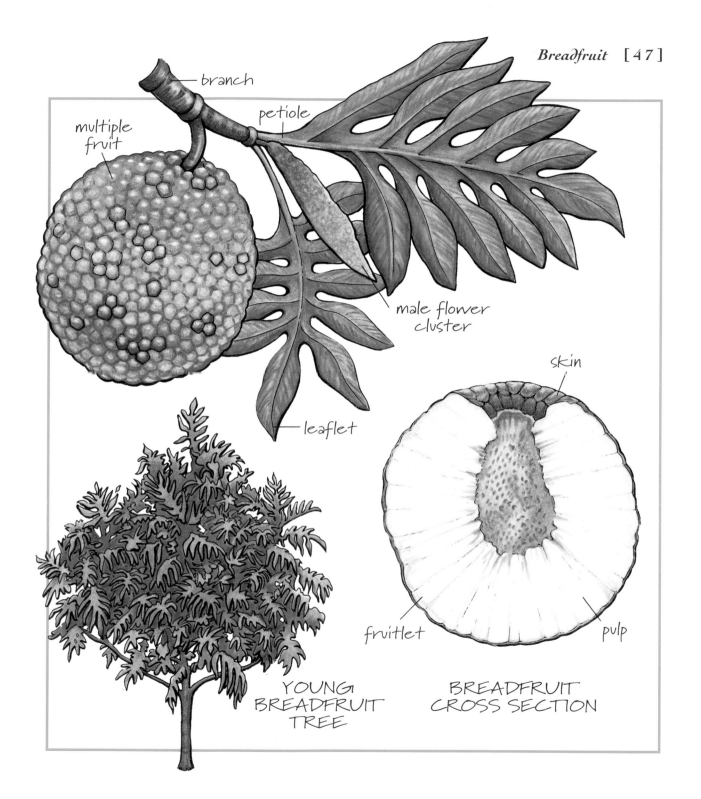

branch

petiole

multiple fruit

male flower cluster

leaflet

skin

fruitlet

pulp

YOUNG
BREADFRUIT
TREE

BREADFRUIT
CROSS SECTION

Let Them Eat Breadfruit

Along with plantains and beans and rice, breadfruit is a starchy staple of the Caribbean and the Pacific Islands. It is not commonly eaten outside these regions, however. Whether sliced and fried like French fries or boiled and mixed with onions, fish, and spices or roasted in oil, breadfruit is a fine and filling food. It is also served raw alongside other fresh fruits such as papayas, mangoes, avocados, and bananas. Sometimes it is mixed with coconut milk to make a pudding. Even its seeds, boiled or roasted, can be eaten.

Not Potato Fruit, nor Rice Fruit, It's Breadfruit!

We don't know exactly who coined the word breadfruit to describe the taste of *Artocarpus communis*. But folks generally agree that French novelist Alexandre Dumas's description from the 1800s is still accurate. Dumas wrote that breadfruit "tasted like the crustless part of fresh bread with a slight suggestion of artichoke plus Jerusalem artichoke." (The Jerusalem artichoke is a member of the sunflower family and grows an edible tuber.) Breadfruit pulp even looks and feels like fresh bread. So there you have it!

Roasted breadfruit is still a popular dish in Hawaii and other southern Pacific islands.

Dig In!

BREADFRUIT DONUTS
(1½ Dozen)

If you live in a large city or in Hawaii, you may be able to find some fresh breadfruit. If so, try this spin on the good old donut. These are hole-free.

1 egg, beaten
1 cup ripe breadfruit, uncooked
2 tablespoons chilled butter
1 cup flour, unsifted
¼ teaspoon nutmeg
⅛ teaspoon cinnamon
¼ teaspoon salt
3 teaspoons baking powder
½ cup sugar
Light oil for frying

Use a fork to cut the butter up into tiny pieces. Mix the butter, egg, and breadfruit. Sift together all the dry ingredients or combine them in a bowl and use a clean fork to stir them together well. Add the dry ingredients to the breadfruit mixture and stir to combine.

Preheat a deep fryer to 375° or heat ½ cup of oil in a heavy frying pan over a high flame. Drop the batter by spoonfuls into the hot oil, being careful not to splash yourself. (You may want to ask an adult for help.) Fry until golden brown, turning them once with tongs or wooden spoons.

Remove the donuts from the hot fat. Drain them on paper towels. If you like, you can put ¼ cup of powdered or granulated sugar in a paper bag and add the donuts. Shake the bag to coat the donuts with the sugar. Enjoy!

Cucumbers
[*Cucumis sativus*]

The phrase "cool as a cucumber" is true! The cucumber's cool, moist crispness is its biggest asset. People testing the inside temperature of a cuke have found it to be as much as 20 degrees cooler than the air that surrounds it.

The cucumber's ancestry dates back almost 9,000 years. It originated in Asia somewhere between Myanmar (formerly Burma) and Thailand. Because of the cuke's antiquity, little is known about who moved it or when. But cucumbers spread easily throughout Asia, taking hold in China as well as in India. Food historians think the fruit vegetable ventured from Asia into the Mediterranean area, where it was enjoyed by both the ancient Hebrews in Egypt and the ancient Romans. Desert dwellers in both the Middle East and Africa have especially appreciated the cool cucumber, which is composed mostly of water.

The cool cucumber has become a mainstay of many home gardens.

A cucumber should be well sliced, dressed with pepper and vinegar, and then thrown out, as good for nothing.

—Samuel Johnson

Name Games and Travels

The cucumber is closely related to the watermelon and to a range of gourds. All these crawling, viny plants look very much alike and often have similar names— so similar that in some ancient writings, we can't tell for sure which plant the author really meant. For example, although the Roman historian Pliny wrote that Tiberius, who ruled the Roman Empire from A.D. 14 to A.D. 37, insisted on cucumbers at his table daily, we cannot say for sure that Tiberius liked cukes and not melons.

Although the cucumber was not well suited to grow beyond the sunny Mediterranean, Romans carried the fruit to other parts of Europe. By the 800s, the cucumber was growing in much of France. It was popular in England by the early 1300s. The cuke won converts somewhat later in Germany and the Netherlands. There the pickling idea struck, and miniature cucumbers called gherkins became popular.

In 1494 Christopher Columbus probably took cucumber seeds from Europe to plant in what later became the Caribbean nation of Haiti. Native American gardeners were raising cukes in modern-day Florida by 1539 and in modern-day Virginia by 1584. Pueblo Indian farmers in what is now New Mexico grew cucumbers and other European crops in the mid-1500s, soon after learning of these plants from Spanish explorers in the region. By the 1630s, the gardens of British colonists in New England included cucumbers. In 1634 William Wood, a British colonist and vegetable chronicler, wrote in *New England's Prospect*, "whatever grows well in England grows as well there, many things being better and larger."

Roman emperor Tiberius

By 1659 Dutch cucumber growers in Brooklyn (later a borough of New York City) were selling their produce to pickle makers. These entrepreneurs cured the pickles in barrels and then sold them from stands on the streets of lower Manhattan, thus beginning the commercial pickle business in the United States.

Cucumber Types

The cucumber fruit develops from flowers on sprawling, green vines. It often has tiny spikes or prickles on its skin, especially at the stem end. Commercial growers, however, have developed a variety of smooth-skinned, commercial **hybrids.**

Dark green, longish cucumbers are most familiar to us. But cukes also come in other colors and shapes. The Armenian cucumber is pale green inside and out, and its skin is easy to eat. The lemon cucumber, common in North America before 1950, is yellow and shaped, well, like a large lemon. The English or hothouse cucumber ranges from around nine inches to two feet in length and is slender and uniform in shape. And then there are small pickling cukes called gherkins or *cornichons* (a French word meaning "small horns"). Newer varieties of cucumbers are seedless and sometimes even burpless. (That is, the cucumbers contain less of a chemical that causes some people to burp.)

A young farmhand in 1949 harvests pickling cucumbers on a machine that carries her between rows. This machine eliminated the need to bend or stoop. These days human labor is not needed to harvest pickling cucumbers.

Fresh or Pickled?

Cucumbers are big business. In the United States, people eat about 10 pounds of cucumbers per person per year. Of that, more than 4 pounds are in the form of pickles!

California leads the nation in growing cucumbers for pickling. Florida is the leader in the fresh cuke arena. While methods of planting and harvesting these two crops differ considerably, all cucumbers require a

To Your Health!

Made up mostly of water, cucumbers don't really have much nutrition to offer. What's more, the best parts of the cuke are the ones many people throw away. Cucumber skins contain folic acid, which helps cells to grow and to divide. (But don't eat the skin of a supermarket cuke—both the grower and the grocer have applied chemicals that aren't good for you.) The seeds are full of vitamin E, which helps to keep tissues healthy and probably prevents cancer. In fact, cucumbers have more vitamin E than any other vegetable in our diet.

warm range of temperatures for good growth. At temperatures below 60 degrees or above 90 degrees, cuke growth tends to slow down.

Both types of cucumbers are grown from seed, which the cucumber itself supplies in abundance—about 17,000 seeds to the pound. Machines called precision seeders lay down the seed, about 2 to 6 inches apart, in rows ranging from 12 to 30 inches apart. Pickling cukes are planted in greater quantity and closer together than cukes headed for the fresh-produce market. This is because pickling cucumbers are usually harvested by machines, which damage a greater percentage of the crop. Pickle growers plant more in anticipation of losing plants at harvesttime.

As the plants develop, they must be fertilized and irrigated. Cukes need about an inch of water per week—more when the weather is extremely hot. After the flowers develop, honeybees play an important role in **pollination.** Cucumber growers bring bee hives to the fields as the plants bloom. Because the plant's flowers are open for just one day, the bees must get in and do their work in a timely fashion so the cucumbers will produce seeds.

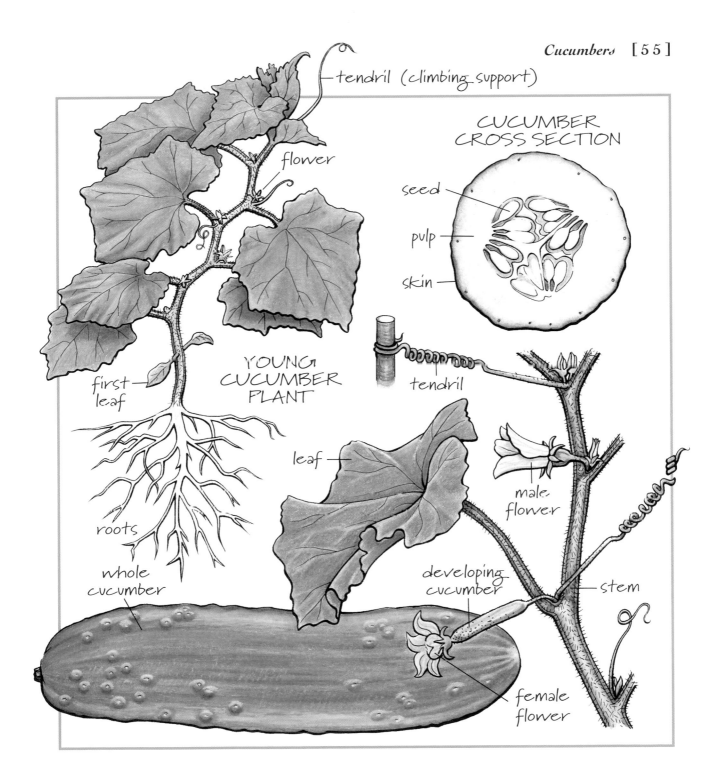

tendril (climbing support)

CUCUMBER CROSS SECTION

flower

seed

pulp

skin

YOUNG CUCUMBER PLANT

tendril

first leaf

leaf

roots

male flower

whole cucumber

developing cucumber

stem

female flower

Some 36 to 50 days after the initial planting, cukes grow to pickable size. In fact, some gain 40 percent of their weight in 24 hours! Farmworkers go into the fields at frequent, regular intervals—once every two or three days—to gather the cucumber fruits. Workers harvest pickling cukes with mechanical pickers but handpick cucumbers for the fresh market. Trucks haul the crop to processing plants, where the cukes are quickly cooled before shipment to supermarkets or to the pickle factory for further processing.

So What Exactly Is a Pickle?

A pickle is a cucumber with an attitude, a cuke that has been soaked in a solution of water and spices. Pickling, or curing fresh vegetables in **brine,** is an ancient method of storing fresh foods for long periods of time. In modern grocery stores, pickles packed in brine have a two- to three-year shelf life.

On a hot day in Virginia, I know nothing more comforting than a fine spiced pickle, brought up trout-like from the sparkling depths of the aromatic jar below the stairs of Aunt Sally's kitchen.
—Thomas Jefferson

If you look closely at these pickles, you can see little bumps on their skins. These bumps are called warts. North American pickle fans expect to find about 7 warts per square inch. Europeans expect to find none!

Most everyone is familiar with sweet and dill pickles. The sweet pickle is prepared with sugar. Dill pickles are tangy and tart. At one time, all dill pickles were prepared with the actual seeds of the dill weed plant. These days it is more common to use the oil of the dill plant in commercial processing.

Eating more than five million pounds of pickles per day, North Americans average about 106 pickles per person per year. Pickles are eaten whole, sliced on sandwiches or burgers, chopped up in relish, and mixed in with other foods like pasta and potato salads.

But we don't always have to pickle cucumbers to enjoy them. Eaten whole as a finger food; as a sauce for fish; as a cooling addition to salads, cold soups, or yogurt; or even as an unusual ingredient in hot soups; cucumbers fill a niche and fill it cheerily.

That's more than you thought you ate, right?

When is a Cuke Not a Cuke?

When it's a . . . .

Sea Cucumber—A real cuke lookalike, this soft-fleshed, prickly sea animal is considered a delicious food in Asia.

Indian Cucumber—This tuberous root is shaped like a cucumber, is eaten raw like a cucumber, and tastes like a cucumber. The Indian cuke is native to northeastern North America. Its botanical name is *Medeola virginica*.

Wild Cucumber—The spiky, edible wild cucumber grows on a vine and is closely related to the domesticated cuke. This plant's botanical name is *Marah oreganus*.

Cucumber Tree—A member of the magnolia family, the bark of this tree has been used to treat malaria (a disease carried by mosquitoes). The tree's fruits resemble small cucumbers but are not edible.

Cucumber Lake—This lake in the Canadian province of Ontario is shaped like—what else?—a bumpy cucumber.

Dig In!

CUCUMBER RAITA
(2 CUPS)

1 small cucumber
1½ cups plain yogurt
1 teaspoon ground cumin
½ teaspoon minced cilantro
2 teaspoons minced mint (optional)
½ teaspoon salt

Peel the cucumber. Cut it in half and scoop out the seeds with a spoon. Grate the firm flesh on the large holes of a kitchen grater. In a bowl, whisk the yogurt until smooth. Add the remaining ingredients. Stir well and chill. Eat with spicy foods for a refreshing treat.

Often served with hot curries in India, raita puts the cool cuke to excellent use.

Because cucumbers cannot survive cold temperatures, growing cukes in greenhouses is necessary to produce the fruit throughout the year.

Greenhouse Cukes

The earliest cucumber farmers probably let their plants wander off and grow at will. The plants attached themselves to weeds and trees the way a wild vine would. But as people learned more about cucumbers, they realized that if a vine were lifted up off the ground, it might produce more fruit, resist insect attacks, and experience fewer diseases.

After forcing cukes to grow up off the ground, it was an easy step to grow them indoors. In greenhouses continuous plantings under controlled conditions—consistent light, warmth, and air circulation—yield fruits over many months. Farmers raise the long, bump-free European or English varieties in greenhouses under methods pioneered by the British and the Dutch. Some methods rely on soil and traditional fertilizers. Others employ hydroponics.

To Your Health!

The cucumber is highly soothing to sunburned, dry, or pimply skin. Cucumber lotions of all kinds are available in specialty stores and natural-food stores. If you've cried your way through a sad video and don't want to appear in public looking like a puffer fish, put a couple of cuke slices on your closed eyelids, lie back, and dream for a few minutes. Remove the slices and head out to meet your public, refreshed and with eyeballs back to normal.

Squashes

[*Cucurbita pepo* — summer]
[*Cucurbita maxima* — winter]

When is a squash a melon? When it's being described by a European who has never seen one before. In about 1540, a scout with the Spanish explorer Francisco Vásquez de Coronado was traveling in what would become the American Southwest (an area that includes Texas, New Mexico, and parts of Arizona and California). The scout reported the region to be full of melons. Around the same time, Jacques Cartier, a French explorer, wrote of *gros melons,* or big melons, in lavish abundance along the St. Lawrence River. These wondrous plants were undoubtedly pumpkins, as there were no melons in the Americas at that time and no squash or pumpkins in Europe.

Sweet Dumply winter squash

Oh Great Pumpkin, you're going to drive me crazy!!!

—Linus, from "Peanuts."

Though related to melons and gourds, both primarily Eastern Hemisphere plants, the squash was American all the way. Many different varieties probably arose in Mexico and Central America. Some squashes, like pumpkins, grow on trailing vines. Others, like zucchini, are bushlike. All produce large, often yellow flowers that must be pollinated by bees. The big, deep-green leaves of the squash plant have toothlike edges and tiny hairs on the surface.

The existence of squashes was no news to native Americans. Of the sacred Indian foods—corn, beans, and squash (with chilies included in the Southwest), squash was the first plant to be cultivated. Native peoples all across North America favored pumpkins, which they roasted in the hot ashes of fires to put into soups. Often hunks or slices of pumpkin were strung up and dried in the sun for later use. The Anasazi who lived in the Southwest from A.D. 100 to A.D. 1300 made pumpkin cakes and fried them on heated flat stones. The later Pueblo people combined pine nuts with pumpkin to make a bread.

As the story of Coronado's scout suggests, the first Europeans to come across squash were the Spanish. In about 1520, a monk traveling with Hernán Cortés's party noted squash seeds and blossoms at the table of the Aztec ruler Montezuma. Another Spaniard, the explorer Álvar Núñez Cabeza de Vaca, reported squash growing near what later became Tampa Bay, Florida, in 1528. The Spanish carried squashes back to Europe in the early 1500s, and European botanists soon began writing about the vegetables. Except for Italy, though, European tables didn't feature squash until the nineteenth century.

The Italians latched onto a skinny, green-skinned squash, laboring over its development and cultivation in the eighteenth century. The Italian name for this vegetable, zucchini, caught on in much of the world. In some places, such as Britain, people call it by the French word *courgette*.

The popular summer squash known as zucchini can be cooked or served raw in salads.

Fig. 1197.—Crook-neck Early Bush Gourd.

Fig. 1200. Turk's Cap Gourd.

Fig. 1201.—White Squash.

Fig. 1198.—Mammoth Pumpkin.

Fig. 1199.—Custard Pumpkin.

Fig. 1202. Patagonian Squash.

By the 1800s, Americans were growing and enjoying several different varieties of squash.

So Many Squash, So Little Time

To speak of "the squash" is meaningless when we consider its multiple varieties from many different locations in the Americas. The majority—winter squashes such as spaghetti, butternut, acorn, and pumpkin—originated in Mexico and Guatemala. There are also Hubbard from the West Indies, cushaw from Florida, and turban from Brazil. Their summer counterparts include zucchini, yellow crookneck, and pattypan or pineapple squash from Chile.

Perhaps because squashes were such everyday foods, little is known about how they were carried beyond Europe to Africa and Asia. We can assume, though, that Dutch and British merchants probably brought them to their colonial outposts in Indonesia, India, and beyond in the 1600s. Arab traders sailing along the African coast must have carried the fruits to west coast sea-ports—such as modern-day Essaouira, Morocco; and Luanda, Angola—at about the same time.

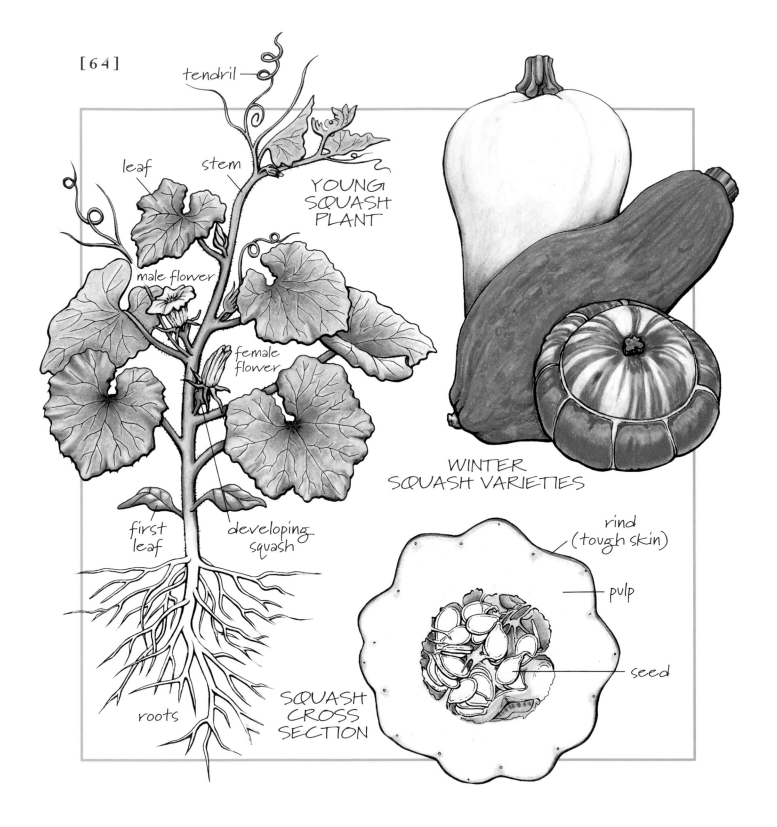

tendril

leaf

stem

YOUNG
SQUASH
PLANT

male flower

female
flower

first
leaf

developing
squash

roots

WINTER
SQUASH VARIETIES

rind
(tough skin)

pulp

seed

SQUASH
CROSS
SECTION

Bees Needed! Growing Squash

Growing squash is very much like growing cucumbers, a close cousin. Workers plant the seeds about 1 to 1½ inches deep in rows 40 to 48 inches apart. Commercial growers use about three to four pounds of seed per acre.

In the early stages of growth, squash plants can be weeded with machines. After squash plants are close to maturity, however, they require hand labor. Workers thin well-established young plants to allow about 8 to 25 inches between each. To allow for the viny growth of larger winter squash, growers thin plants to about one per 3 to 5 feet, with 6 to 8 feet between rows.

Like cucumbers, squash plants need bees to pollinate their large, yellow, male and female flowers. Growers generally place one hive per acre in the fields. Once the fruit begins to grow, plants must be harvested two to three times per week. Pickers pluck off summer squash varieties when they are about 1 to 1½ inches in diameter. Winter squash is ready to pick when its skin has fully hardened.

Gathered into nylon or canvas field bags, the squash moves quickly to the processing area. There the fruits must be cooled down before laborers sort and pack the squash into cartons for truck transport.

Like its cousin the cucumber, the squash plant grows a large yellow flower *(above)* that needs to be pollinated. These yellow crookneck squashes *(left)* can be picked as a summer variety while they are immature and their skin is still soft.

Green scallop squash

Going for Broke(r)

Like most of the food we eat, squash must get from the farm to the store so the customer can buy it. The people who make this happen are called brokers. Brokers buy from farmers and sell to retail outlets. Sometimes brokers will hold exclusive contracts with particular growers, year after year, for certain crops. Other times brokers will simply call up farms, see what they have available, and then sell the produce directly from field to supermarket chain. Brokers often spend hours each day on the telephone seeking vegetables to sell, especially in times of crisis, such as when a late spring freeze wipes out an entire crop of acorn squash. Supermarkets will be calling the brokers to keep shoppers supplied, regardless of the shortage.

Wholesalers run a type of brokerage operation, too. They contract with growers, repack the produce in their own warehouses, and sell it themselves under their own brand names.

The Whole Squash

Squashes provide edibles from start to finish. Their blossoms are tasty dipped in flour and fried, and their roasted seeds make a satisfying snack. Modern Mexicans eat squash this way, as they did in ancient times. Russians and Middle Easterners have picked up on these foods, too. And in Greece at springtime, people serve a country dish called *kolokithokorfades,* or zucchini tops. People sauté the blossoms in butter with chopped green onions, then top the food with tomato sauce and grated cheese.

Besides eating the flowers and seeds, squash eaters enjoy soups made from a combination of squashes cooked in broth and then topped off with grated cheese. In North America, where squashes of all kinds have been eaten for thousands of years, the zucchini has become an emblem of summer garden excess. By August most home gardeners have more than they can use, and some even leave bushels of zucchini on strangers' doorsteps, desperate to share the wealth. The number of North American zucchini recipes seems endless. Zucchini bread, zucchini soup, zucchini pancakes, zucchini omelets, zucchini pasta, zucchini pizza, and even zucchini preserves are popular.

And in the fall at pumpkin time, family cooks concoct delicious soups, puddings, pies, muffins, and breads. Some people like to cut butternut squash in half, scoop out the seeds, and bake the halves with a dollop of molasses and butter in the cavities. Spaghetti squash is often served baked or boiled with tomato sauce and cheese, as might be guessed. All winter squashes make good baking dishes—just stuff them with a precooked, spicy meat and rice mixture and bake until the squash is soft.

Pumpkin bread and zucchini bread are just two of the delicious recipes that use squash.

Carve That Pun'kin

In October the air is crisp and pumpkin plants sprawl in the fields. The pumpkin becomes North America's designated squash, beating out the zucchini across the board. An autumn ritual is at hand: selecting a pumpkin and carving a jack-o'-lantern.

Making vegetables into portable lamps is not new. In ancient Britain, people hollowed and carved large turnips into lanterns as part of their New Year celebration on October 31. Although the celebration took on new names and meanings over the centuries, the lantern-carving tradition lingered on even into the 1600s, when British and Scottish

Jack-o'-lanterns are a common sight when autumn comes around.

immigrants brought the practice to North America. The newcomers quickly realized that pumpkins were not only abundant in the fall but were easier to hollow out than turnips. The rest is Halloween history.

Modern jack-o'-lanterns range from homemade carvings cut out with a dull kitchen knife to elaborate patterns sculpted by professional artists with precision tools. Around Halloween images of pumpkins, carved and uncarved, decorate fabrics, toys, and even greeting cards—strong evidence of our affection for this always tasty, sometimes spooky fruit vegetable.

To Your Health!

Winter squashes, such as acorn and spaghetti, are the nutritional powerhouses of the vegetable world, high in potassium and full of iron and protein. Summer varieties have less nutrition but please in other ways as versatile and colorful additions to soups and stir fries, or even as pizza toppings.

Dig In!

ZUCCHINI SALAD
(4–6 Servings)

1 pound zucchini, sliced
4 to 6 tablespoons olive oil
2 to 3 tablespoons lemon juice
2 cloves garlic, crushed
1 teaspoon caraway seeds, crushed with
   a mortar and pestle or chopped
   coarsely
pinch of paprika
salt and pepper to taste

Pour about ½ inch of water into a medium saucepan and bring to a boil. Place the squash in a steaming basket, set the basket in the pan, cover, and cook until tender, about 4 to 8 minutes. (If you don't have a steaming basket, then use half the water, put the squash directly into the pan, cover, and cook.) Drain the squash. Combine the oil and the lemon juice in a serving dish. Then add the garlic, caraway seeds, salt, and pepper. Mix well. Add the zucchini and gently toss all the ingredients. Top with paprika and serve hot or cold.

This easy-to-prepare Tunisian salad features crushed caraway seeds and aromatic herbs closely related to parsley.

Peppers
[Capsicum annuum — sweet peppers]
[Capsicum frutescans — hot peppers]

Chili-heads and pepper poppers are everywhere! People throughout the world eat peppers in infinite variety of color, shape, heat content, and, of course, taste. Pepper plants can be tall or short, their fruit sweet or hot, their pods any shape at all from round to pencil thin, and most any color from yellow to green to purple. But generally their flowers are the typical solanum family star shape, and their leaves are bright green and quite uniformly spade-shaped. These remarkable plants provide not only fresh fruit but also the dried powders that make the pepper family the most-used spice in the world. While the world eats peppers both hot and sweet, the cultural impact and culinary importance of hot peppers far outweighs that of the sweet.

An assortment of hot peppers. The red habanero peppers in the center are the hottest of the hot peppers.

Peter Piper pick'd a peck of pickled peppers. Where is the peck of pickled peppers Peter Piper pick'd?

—J.K. Paulding

A little pepper burns a big man's mouth.

—Jamaican proverb

First Farmers

The pepper plant may have originated many thousands of years ago in what would become Bolivia. The plant was a good traveler. Its seeds could last a long time and were easily carried both by birds and people. From its home territory in Bolivia, the pepper plant was rapidly and widely distributed in Mexico, Central America, and the Caribbean. Wherever peppers went, they took hold much as does a weed, fitting into whatever conditions they found.

The best evidence of the consumption of wild peppers comes from the area of Mexico City where scientists have dated pepper seeds 9,000 years old. The domesticated pepper was probably under cultivation about 2,000 years later. We know two things about early pepper growing. One is that farmers sowed the seeds individually, much as they do in modern times. This method differs from the early European style of tossing out handfuls of seed into a prepared field, a technique known as "broadcast" sowing. The second thing we know is that pepper fields were irrigated by water run through trenches.

One of the first cultivated plants in the Americas, along with beans and squash, the pepper was improved by human involvement. For example, the domestic pepper's fruit hangs down, in contrast to that of wild varieties that depend on birds for dispersal of seeds. Such wild peppers grow their fruit pointing upward, where birds can find it. Hanging fruits suggest that the earliest farmers encouraged those varieties that eluded the ever-hungry birds. This arrangement also allowed the plant to support heavier, larger fruit more worth the picking.

It's a Fact!

For early Americans, chilies could be a potent weapon. Burning hot peppers makes a choking, eye-stinging smoke with which the Native Americans repelled invaders. Warriors also daubed the hottest chili essence on the tips of their weapons. Weapons thus treated could make injuries more painful and disabling.

Some pepper plants, like these colorful Hawaiian hot peppers, grow their fruit pointing upward.

The early people of Mexico and Central America—the Maya, the Aztecs, and the Toltecs—and, later, the Indians of the American Southwest all grew peppers, most frequently the hotter varieties. They ate their peppers with everything—stewed with meat; frothed with chocolate in a warm, unsweetened drink; mixed with squash in soups; sprinkled on all manner of tortilla and burrito concoctions; and boiled with corn and potatoes. The chili pepper has appeared in ancient carvings, on pottery, and in tapestries, and has even functioned as a form of currency.

The Wayfaring Pepper

But peppers reached beyond the Western Hemisphere when Christopher Columbus arrived from Europe in search of spices. Carried back to Spain in 1493 by members of the Columbus expedition, sweet and hot pepper plants began to make their global moves. Traders from Portugal, a country bordering Spain, had brought peppers to East Africa, India, Indonesia, and elsewhere in Asia by the early 1500s.

Peppers were known in Italy by 1535 and in Germany by 1542. Hungary, however, is the one European country always identified with chili peppers, which the Turks probably introduced in the mid-fifteenth century. The Turks may have met up with chilies through the Italians or possibly through trade with

Portuguese colonies in India. In any event, peppers appealed to the Hungarian people, and soon they were hanging chilies from the roof beams and drying them on the rooftops, just as in other pepper-mad parts of the world.

But peppers, both sweet and hot, reached well beyond Europe. China received them from India or perhaps Thailand, again courtesy of Portuguese traders. By the 1600s, pepper plants were established throughout Asia.

Back in the Western Hemisphere, the pepper plant marched northward from Mexico with the Spanish early in the history of the American Southwest. Juan de

Many famous sea voyages were attempts to find shortcuts to spice trade centers. Europeans craved spices to improve their bland diets and to hide the taste of spoiled meat.

It's a Fact!

Confusion between pepper, the Asian spice, and peppers, the American fruit, stems from 1492 when Columbus first saw the Arawak Indians of the Caribbean munching on what they called axi. Still absorbed in the delusion that he had found "the Indies," Columbus could not fully comprehend that he was looking at a plant altogether new to him. So he dubbed the plant *pimiento*, or pepper plant, after the real pepper plant from India, known in Spanish as *pimienta*.

Oñate, founder of Santa Fe, New Mexico, is often credited with bringing peppers to what became the United States in 1583. But it seems likely, too, that the Pueblo Indians had long before received varieties of chilies through trade with native Mexicans. For thousands of years, Pueblo people had been using a wild chili, known as the *chiltepín*, a tiny-berried plant still growing wild in parts of New Mexico, Texas, and southern Arizona.

It wasn't until 1621 that the two kinds of pepper plants finally reached the English colony of Virginia, arriving with other food-stuffs as a gift from the governor of Bermuda (a Caribbean island group). George Washington grew peppers at Mount Vernon, as did Thomas Jefferson at Monticello. According to pepper historian Jean Andrews, Jefferson wrote in 1812 that he had received capsicum seeds from "a traveler to Taxas," that is, Texas, then a part of Mexico.

Cooking with Capsicums

The Southwestern chili dish *posole* is a hominy corn and pork stew with a chili and garlic sauce served on the side and garnished with chopped onions and cilantro. And of course, there's chili con carne, especially beloved in Texas. Another chili dish, common in New Mexico, is *huevos rancheros*, eggs poached in green or red chili sauce and served over warmed corn tortillas. The mild bell pepper is also appreciated in the South-west, well seasoned and filled with rice and cooked meat.

A chili cuisine called Cajun flourishes in Louisiana, where people cook red beans and rice with sausage, onions, and chilies and then spice the concoction with chili-based Tabasco sauce. The dish has long been a staple, along with crayfish *étouffée*—a combination of green onions, Tabasco sauce, celery, and the local shellfish.

And who can think of Hungarian food without paprika! In Hungary paprika refers both to the ground powder and to the pepper fruits themselves, whether sweet or hot. Combining fat, onions, and ground peppers, Hungarians have created a memorable cuisine. Goulash, which is really a soup, combines meat, onions, and paprika served over small dumplings and is often garnished with sour cream.

Chili peppers have also been hot items in the cuisine of India, long known for its spicy food. Brought by the Portuguese, who took over the coastal city of Goa in the early 1500s, chilies grew throughout the subcontinent with ease. They were quickly added to *masala*, India's basic ground mix of spices. But peppers also stood deliciously alone, appreciated in a country with millions of vegetarians. Cooks prepare sweet or hot peppers in a pungent sauce or include them in other foods—from *samosas* (deep-fried dough filled with seasoned vegetables) to spicy dishes

Southwestern nachos—which include corn chips, refried beans, salsa, and cheese—are topped with sliced jalapeno peppers.

It's a Fact!

Amal Naj, the Indian-born author of a book called *Peppers*, declares that many Indian dishes contain the chili pepper used three ways—dry pods cooked in oil, then the ground pepper, and finally fresh hot peppers.

called curries. Even the traditional condiments known as chutneys (pickled combinations of vegetables and fruits) welcomed both chili and sweet peppers.

China, with its varied cuisines and people willing to try anything, has been described as a place waiting for the chili pepper. And the country not only eats the largest total amount of chili peppers but also grows, processes, and exports the most, too. Peppers are most closely associated with two provinces, Sichuan and Hunan, in central China. The starting ingredient for hot Chinese food is an oily paste of chilies, garlic, and sesame seeds. People add spring onions and bits of anything and everything else—shrimp, vegetables, chicken, beef, and so on. Breakfast in Sichuan is often plain noodles with chili pepper oil lavishly tossed on top.

Thailand is another country full of chili eaters. Introduced to the plant by foreign traders, the Thais love chilies so much that minced "Thai" chilies in fish sauce appear on every dining table. According to pepper chronicler Amal Naj, Thais eat more hot peppers per person than any other nation in the world. *Pad Thai*, the national stir-fry noodle dish, combines shrimp, tofu, and hot pepper flakes with peanuts and, of course, noodles.

Curries are another key part of Thai cuisine. People migrating from India first brought curry, which the use of particular Thai ingredients helped to alter. One Thai-style curry dish is green curry, made with the hottest green chilies and often cooked with pork, coconut milk, and fresh basil.

As in other parts of the world that have embraced the heat of chilies, African cooking starts with bland but filling fare—such as cassava, millet, couscous, or rice—and adds hot spicy sauces, sometimes containing meat, sometimes just vegetables. A basic sauce from Ghana known as *chilli sambal* mixes or mashes fresh red chilies together with onions, salt, and tomatoes. An Ethiopian spice mix called Berbere includes both cayenne (one of the hottest chilies) and paprika, along with cumin, coriander, ginger, cardamom, fenugreek seeds, nutmeg, cinnamon, cloves, onion powder, allspice, and black pepper. (Doesn't leave much out, does it?) A vegetable stew called *wot* uses Berbere to season corn, pumpkin, and plantain, a bananalike fruit.

Pimento Power

Ever wonder about that red stuffing in green olives? Usually it's ground-up pimentos, a heart-shaped variety of sweet red pepper. The pimento, although originally from the Americas, was early on cultivated in Spain. Americans imported their pimentos until 1911, when farmers in the state of Georgia managed to grow the peppers. Three years later, a Georgian inventor built the first pepper-roasting machine. The American pimento-canning industry was born. Oddly enough, U.S. consumers can rarely find fresh pimentos. But these peppers turn up regularly in casseroles, cheese sauces, and spreads—and, of course, squirted into the hearts of green olives.

Dried red chili peppers are a common sight at markets in China.

To Your Health!

Perhaps because of their heat, chilies are good for health. Chilies have a stimulating effect on the digestive system and open clogged sinuses easily. Capsaicin creams have worked wonders on people suffering from arthritis and rheumatism. Headaches of all sorts respond well to capsaicin administered through nasal sprays. Chilies may also work to lower cholesterol and to prevent heart attacks. And we haven't even mentioned vitamins. All kinds of peppers are high in vitamin C, which keeps tissues healthy and promotes healing, and vitamin A, important to the eyes and the skin.

Call the Fire Department!

Most of the heat in a pepper is contained in the seeds and veins of the pod. Smaller means hotter, since small chilies contain the same number of seeds as the largest of sweet bell peppers. If you crunch down on a chili that's too hot to handle, don't reach for a soda or a glass of water. Both will simply spread the fire around. Ask for bread, milk, or ice cream—or, best yet, *sopapillas* with honey, the deep-fried, puffy sweet breads that accompany many Mexican and New Mexican meals.

What makes a chili hot? It's capsaicin, a chemical that not only causes a burning sensation, but also triggers endorphin, a natural painkiller we carry around inside us without need of a doctor's prescription.

When a person bites into a hot pepper, the capsaicin makes the tongue feel on fire. It really isn't, of course,

The seeds and veins of the chili pepper contain the chemical that make the pepper so hot. It's a good idea to wear gloves when handling hot peppers, because the seeds can burn your skin.

but the brain doesn't know that. The brain feels real danger and so tells the body to go into defensive action. The skin breaks out in a sweat, the mouth fills with saliva, and the nose instantly requires blowing. And the brain, still fooled, releases endorphin, which helps us cope if we are indeed in physical pain. In this case, the pain is no real threat to our well-being, so the endorphin makes us feel great. We take another bite of chili pepper and get another dose of endorphin.

Crazy for Chilies

The story of the chili pepper in the United States focuses on two areas, Louisiana and the Southwest. Each fall in most every town in New Mexico, people buy green chilies in huge sacks and either roast them themselves or watch as local sellers toss the chilies into rotating metal bins slowly turning over a gas flame. Others buy red chilies and make them into *ristras*, the long strings of chilies tied together to hang by the front door or in the kitchen. New Mexico is the country's leading producer of chilies, and in fact, chilies account for the majority of the state's fruits and vegetables. New Mexico is also the home of the Fiery Foods Show, an annual event in Albuquerque that hosts more than 200 vendors. They sell every red and green chili combo imaginable—hot sauces, salsas, pestos, pastas, mustards, beers, barbecue sauces, spicy snacks, and jellies. The state's

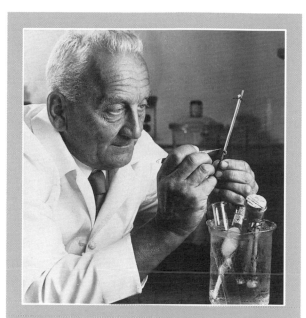

The Search for C

Peppers were the plant from which Hungarian Nobel Prize winner Albert Szent-Gyorgyi finally extracted sufficient quantities of vitamin C with ease after trying other sources for years. The story goes that his wife served him some chilies for dinner, as she undoubtedly had before. But this dinnertime, he looked at his plate as a scientist rather than as a diner and immediately took them off to his lab. He won the prize for medicine and physiology in 1937.

Dried chili ristras hang from a building in Santa Fe, New Mexico.

chili-related business, all told, amounts to more than $2.5 billion a year in sales.

Santa Fe, the capital of New Mexico, boasts the finest in Southwest cuisine and has sent forth its chefs to reproduce that style and taste throughout the world. Las Cruces in southern New Mexico is the home of New Mexico State University, the academic center of chili research. And the towns between Las Cruces and Truth or Consequences live from raising and harvesting chilies.

But what about Louisiana? Home of Tabasco sauce! The story goes that U.S. soldiers fresh from the Mexican War arrived in New Orleans in 1847 to recuperate. They brought chili peppers of the Tabasco variety with them. A man named Maunsell White grew the first Tabasco peppers on his plantation and shared some seeds with Edmund McIlhenny, who turned the Tabasco seeds into an empire of hot sauce poured into tiny bottles. By 1872 McIlhenny was selling his sauce in Europe as well as marketing it throughout the United States.

It's a Fact!

Almost all the chilies used in McIlhenny's Tabasco Pepper Sauce are grown in Latin America and shipped to Louisiana as pepper "mash" in barrels. Only about 10 percent of the year's crop is still planted at McIlhenny's headquarters on Avery Island, Louisiana.

Dig In!

RATATOUILLE CREOLE
(4–6 Servings)

¾ cup olive oil

2 onions, peeled and sliced into rings

2 each, sweet red and green peppers, seeded and sliced into thick strips

1 pound zucchini, chopped

2 large cucumbers, peeled and thickly sliced

2 medium eggplants, peeled and chopped

1 pound tomatoes, peeled and thickly sliced, or one large can of tomatoes

dash of salt and pepper

1 teaspoon mixed herbs, including thyme, bay, parsley and chives. (Note: If it's easier, use a prepared herb mix that contains at least some of these ingredients.)

sugar to taste (optional)

Preheat oven to 200° F. Pour the olive oil into a very large ovenproof dish, add the onions, and bake uncovered until the onions soften. Remove the dish from the oven and arrange the peppers and eggplant on the bottom, and then put in the zucchini. Layer the tomatoes and cucumbers on top. (Use a fork to handle the vegetables, as the dish will be hot.) Add salt and pepper to taste. Cover and bake 10 minutes, then uncover and add the herbs and the sugar. Continue cooking until all the ingredients are cooked and much of the juice absorbed. The dish should be moist, not dry, however! Serve hot. (Adapted from *A Taste of Africa* by Dorinda Hafner.)

This recipe comes from the West Indies islands of Martinique and Guadeloupe, areas peopled originally by Native Americans and then by Europeans and Africans.

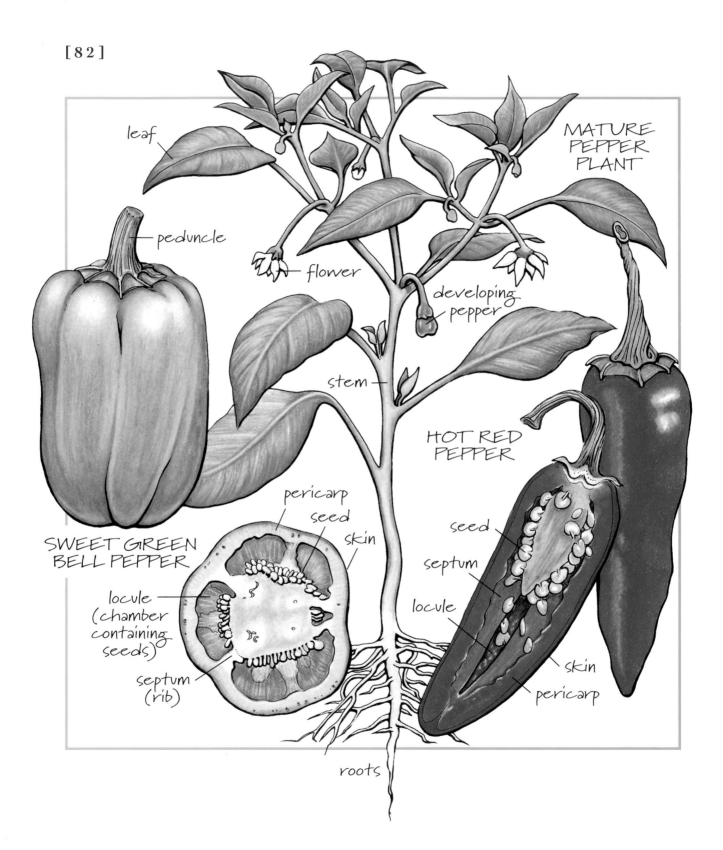

leaf

peduncle

flower

MATURE
PEPPER
PLANT

developing
pepper

stem

SWEET GREEN
BELL PEPPER

pericarp

seed

skin

HOT RED
PEPPER

seed

septum

locule

skin

pericarp

locule
(chamber
containing
seeds)

septum
(rib)

roots

The Techno Pepper

The pepper plant is used as a coloring agent in prepared meats, cosmetics, desserts, cheeses, and more. Commercial chicken raisers feed their birds pepper extracts so that egg yolks will have that jolt of yellow and so that the chickens, once slaughtered, will reveal yellow fat, not pasty white. The capsaicin from hot peppers turns up in ointments used by athletes for sore muscles. In full strength, capsaicin appears in spray products that repel dogs, insects, and sometimes, unappreciated people.

Planting and Picking Peppers

Farmers raise peppers largely with hand labor. Some growers start their seeds outside, using special hand-powered or mechanical planting tools to deposit each seed. Other growers start pepper seeds in greenhouse beds. When the seedlings are six to eight inches high, they are ready to go outside. Workers prepare the fields for planting and cover the rows with black plastic mulch. Then field hands punch holes in the plastic and lower the young plants into place.

Pepper plants need frequent watering during their growing cycle. After about 70 to 90 days, the fruit is

A field of wax peppers grows in Florida. Chili pepper plants are commonly harvested three times—twice when still green and once when the pods have ripened to red.

Laborers in California dump harvested red bell peppers into field bins.

ready to harvest. Even though mechanical harvesting is gaining popularity, most farmers harvest the tender and easily bruised fruits by hand. Crews of pickers move slowly down the rows cutting the pods off the plants with clippers and gathering the peppers into buckets. The workers empty their pails into truck trailers, which take the peppers to processing areas. Laborers then wash, sort, and grade the peppers. The fruit vegetables next travel to supermarkets, where they'll be sold fresh, or to processing plants, where they might be canned, bottled, pickled, or made into any number of pleasing pepper products.

Glossary

brine: A concentrated mixture of salt and water used to pickle foods.

domestication: Taming animals or adapting plants so they can safely live with or be eaten by humans.

dormant: Not growing, but capable of growing later.

heirloom garden: A garden in which traditional or historical, but commercially unimportant, varieties of plants are grown.

hybrid: The offspring of a pair of plants or animals of different species.

pesticides: Poisons that growers apply to crops in order to kill unwanted insects or weeds. Pesticides can also harm humans and animals.

photosynthesis: The chemical process by which green plants make energy-producing carbohydrates. The process involves the reaction of sunlight to carbon dioxide, water, and nutrients within plant tissues.

pollination: The placement of pollen on a flower so that fruit will grow from the blossom. Bees pollinate the flowers of many plants.

staple food: A food plant that is widely cultivated across a given region and used on a regular basis.

tropics: The hot, wet zone around the earth's equator between the Tropic of Cancer and the Tropic of Capricorn.

Further Reading

Hill, Lee Sullivan. *Farms Feed the World.* Minneapolis: Carolrhoda Books, 1997.

Inglis, Jane. *Proteins.* Minneapolis: Carolrhoda Books, 1993.

Johnson, Sylvia A. *Tomatoes, Potatoes, Corn, and Beans: How the Foods of the Americas Changed Eating around the World.* New York: Atheneum Books for Young Readers, 1997.

Nottridge, Rhoda. *Vitamins.* Minneapolis: Carolrhoda Books, 1993.

Root, Waverley. *Food.* New York: Simon & Schuster, 1980.

Trager, James. *The Food Chronology.* New York: Henry Holt and Company, 1995.

Vegetarian Cooking around the World. Minneapolis: Lerner Publications, 1992.

Wake, Susan. *Vegetables.* Minneapolis: Carolrhoda Books, 1990.

Colors abound at a marketplace in Guadalajara, Mexico.

Index

About the Author

Meredith Sayles Hughes has been writing about food since the mid-1970s, when she and her husband, Tom Hughes, founded The Potato Museum in Brussels, Belgium. She has worked on two major exhibitions about food, one for the Smithsonian and one for the National Museum of Science and Technology in Ottawa, Ontario. Author of several articles on food history, Meredith has collaborated with Tom Hughes on a range of programs, lectures, workshops, and teacher training sessions, as well as *The Great Potato Book*. The Hugheses do exhibits and programs as The FOOD Museum in Albuquerque, New Mexico, where they live with their son, Gulliver.

Acknowledgments

For photographs and artwork: Steve Brosnahan, p. 5; TN State Museum, detail of a painting by Carlyle Vrello, p. 7; © Ed Young/AGStockUSA, pp. 11, 66, 84, 86; Corbis-Bettmann, pp. 12, 26, 63; © Grace Davies, pp. 13, 15, 37 (bottom), 40, 41, 56 (bottom), 67, 68, 78; UPI/Corbis-Bettmann, pp. 14, 53, 79; Inga Spence/Tom Stack & Associates, pp. 17,18, 84; © Craig D. Wood, pp. 19, 27, 71; © Walter Pietrowicz/September 8th Stock, pp. 21, 22, 23, 33, 58, 69, 81; © Mark Gibson/AGStockUSA, p. 25; © Frank S. Balthis, p. 29, 65 (top); © Dwight Kuhn, pp. 30, 51, 83 (top); © David Solzberg/AGStockUSA, p. 35; © Pat & Chuck Blackley, p. 36; © Jim Jernigan/AGStockUSA, p. 37 (top); FAO, pp. 38, 59; © Douglas Peebles/ AGStockUSA, p. 43; Corbis, p. 44; © Karlene V. Schwartz, p. 46; © Robert Fried, pp. 48, 76, 77; Archive Photos, pp. 52, 74; Bowdoin College Museum of Art, Brunswick, Maine, Bequest of James Bowdoin III, p. 56 (top);© John Marshall/AGStockUSA, p. 61; © Tony Hertz/AGStockUSA, pp. 62, 65 (bottom); © G. Brad Lewis/AGStockUSA, p. 73; © Brian Parker/ Tom Stack & Associates, p. 80; © Bill Barksdale/AGStockUSA, p. 83 (bottom). Sidebar and back cover artwork by John Erste. All other artwork by Laura Westlund.Cover photo by Steve Foley and Rena Dehler.
For quoted material: p. 4, M. F. K. Fisher, *The Art Of Eating* (New York: Macmillan General Reference, 1990); p. 10, Elizabeth David, "An Omelett and a Glass of Wine," 1984, *Consuming Passions*, Jonathon Green, ed. (New York: Fawcett Columbine, 1985); p. 24, Sophie Coe, *America's First Cuisines* (Austin: University of Texas Press, 1994); p. 34, John Gerard, *The Herball, or Generall Historie of Plantes* (London: John Norton, 1597); p. 42, as mentioned by Waverly Root, *Food* (New York: Simon & Schuster, 1980); p. 50, as quoted by Michael Cader, ed., *Eat These Words: A Delicious Collection of Fat-Free Food for Thought* (New York: HarperCollins, 1991); p. 56, as quoted by Mt. Olive Pickle Company, www.mtolivepickles.com/history.htm (Mt. Olive, NC: 1996); p. 60, Charles Schultz, "Peanuts" (United Features Syndicate, 1961); p. 70, James Kirke Paulding, *Konigsmarke, the Long Finne: A Story of the New World* (London: Cox and Baylis, 1823).
For recipes (some slightly adapted for kids): p. 23, adapted by Meredith Sayles Hughes from a description of the dish by Maguelonne Toussaint-Samat, *A History of Food* (Cambridge, MA: Blackwell Reference, 1992); p. 33, Meredith Sayles Hughes; p. 40, Suad Amari, *Cooking the Lebanese Way* (Minneapolis: Lerner Publications Company, 1986); p. 49, Dorothy Horn, *More Real Guamanian Recipes* (Guam: Pacific Color Press, 1994); p. 58, reprinted with permission from *Cooking at the Natural Café* by Lynn Walters. © 1992. Published by The Crossing Press: Freedom, CA.; p. 69, reprinted with permission from *The World in Your Kitchen* by Troth Wells. © 1993. Published by The Crossing Press: Freedom, CA.; p. 81, Dorinda Hafner, *A Taste of Africa* (Berkeley: Ten Speed Press, 1993).